ITAR AND EXPORT CONTROLS FUNDAMENTALS

A Guide for Compliance Managers

Navigating the Complexities of International Traffic in Arms Regulations

BY

CARL B. JOHNSON

ITAR AND EXPORT CONTROLS FUNDAMENTALS

A Guide for Compliance Managers

Navigating the Complexities of International Traffic in Arms Regulations

Printed in the United States of America

Contents

About the Author

Carl B. Johnson, President of Cleared Systems, is a highly experienced and certified ITAR, CMMC 2.0, Microsoft GCC High, and Microsoft DLP/AIP consultant. With over twenty years of experience in information assurance, cybersecurity, policy development, risk management, and regulatory compliance, he brings a wealth of knowledge and expertise to his clients.

Johnson's passion for technology has driven him to stay on the cutting edge of security solutions and best practices. He has worked with a variety of clients across different industries, ranging from healthcare to banking to aerospace engineering, providing them with expertise on securing their data against external threats.

With a deep understanding of industry standards such as NIST & ISO 27001, Johnson is able to develop comprehensive security solutions that meet both internal customer needs and external customer requirements. His commitment to staying up-to-date with the latest security solutions and best practices ensures that his clients receive the best possible service.

Whether a client is looking to secure their data or meet regulatory compliance requirements, Johnson's experience and expertise make him a valuable asset in providing secure and efficient solutions.

Introduction

Purpose of the Book

The purpose of this book is to provide compliance managers with a comprehensive understanding of the International Traffic in Arms Regulations (ITAR) and export control regulations, and to assist them in navigating the complexities of these regulations in order to achieve ITAR and export control compliance.

ITAR regulations govern the export of defense-related items, services, and technology, and are designed to protect U.S. national security interests and foreign policy objectives. Export control regulations, on the other hand, govern the export of items, services, and technology that have both civilian and military applications and are designed to promote international security and stability.

For compliance managers in the defense industry, understanding and complying with ITAR and export control regulations is a critical responsibility. ITAR and export control compliance is a complex and challenging process that requires a thorough understanding of the regulations and their requirements, as well as the development and implementation of effective ITAR and export control compliance programs.

This book provides compliance managers with the knowledge and tools they need to understand ITAR and export control regulations, to develop and implement effective ITAR and export control compliance programs, and to achieve ITAR and export control compliance.

Overview of ITAR Regulations

The International Traffic in Arms Regulations (ITAR) are a set of regulations administered by the U.S. Department of State, Directorate of Defense Trade Controls, that govern the export of defense-related items, services, and technology. The purpose of ITAR is to protect U.S. national security interests and foreign policy objectives by controlling the export of defense-related items, services, and technology.

ITAR applies to any person who engages in the manufacture, export, import, transfer, or brokering of defense-related items, services, and technology. ITAR-controlled items, services, and technology are listed on the U.S. Munitions List (USML), which is located in 22 CFR Part 121 of the ITAR regulations.

ITAR licensing requirements are determined based on the nature of the item, service, or technology being exported, the end-use of the item, service, or technology, and the end-user of the item, service, or technology. ITAR licensing requirements may also be influenced by the presence of foreign nationals or foreign-origin items in the manufacturing or research and development process.

ITAR recordkeeping and reporting requirements are extensive and include maintaining detailed records of all ITAR-controlled exports and imports, including the date of shipment, the name and address of the recipient, the value of the item, service, or technology, and a description of the item, service, or technology. ITAR also requires the submission of annual reports to the U.S. Department of State, Directorate of Defense Trade Controls, detailing all ITAR-controlled exports and imports.

ITAR compliance obligations are the responsibility of all parties involved in the manufacture, export, import, transfer, or brokering of ITAR-controlled items, services, and technology. Compliance with ITAR is mandatory, and failure to comply with ITAR requirements can result in significant penalties, including fines and imprisonment.

In review, ITAR regulations are a complex and challenging set of regulations that must be understood and complied with by all parties involved in the manufacture, export, import, transfer, or brokering of defense-related items, services, and technology. Compliance managers in the defense industry must have a thorough understanding of ITAR regulations and their requirements in order to achieve ITAR compliance.

Understanding ITAR Regulations

The International Traffic in Arms Regulations (ITAR) are a complex set of regulations that govern the export of defense-related items, services, and technology. Compliance with ITAR is mandatory, and failure to comply with ITAR requirements can result in significant penalties, including fines and imprisonment.

ITAR applies to any person who engages in the manufacture, export, import, transfer, or brokering of defense-related items, services, and technology. ITAR-controlled items, services, and technology are listed on the U.S. Munitions List (USML), which is located in 22 CFR Part 121 of the ITAR regulations.

ITAR licensing requirements are determined based on the nature of the item, service, or technology being exported, the end-use of the item, service, or technology, and the end-user of the item, service, or technology. ITAR licensing requirements may also be influenced by the presence of foreign nationals or foreign-origin items in the manufacturing or research and development process.

ITAR recordkeeping and reporting requirements are extensive and include maintaining detailed records of all ITAR-controlled exports and imports, including the date of shipment, the name and address of the recipient, the value of the item, service, or technology, and a description of the item, service, or technology. ITAR also requires the submission of annual reports to the U.S. Department of State, Directorate of Defense Trade Controls, detailing all ITAR-controlled exports and imports.

ITAR compliance obligations are the responsibility of all parties involved in the manufacture, export, import, transfer, or brokering of ITAR-controlled items, services, and technology. Compliance managers in the defense industry must have a thorough understanding of ITAR regulations and their requirements in order to achieve ITAR compliance.

Use Cases:

A defense contractor is engaged in the manufacture of ITAR-controlled items, services, and technology. The defense contractor must comply with ITAR regulations by obtaining the necessary licenses, maintaining detailed records, and submitting annual reports to the U.S. Department of State, Directorate of Defense Trade Controls.

A defense contractor is exporting ITAR-controlled items, services, and technology to a foreign end-user. The defense contractor must comply with ITAR regulations by obtaining the necessary licenses, maintaining detailed records, and submitting annual reports to the U.S. Department of State, Directorate of Defense Trade Controls.

A defense contractor is transferring ITAR-controlled items, services, and technology to a foreign subsidiary. The defense contractor must comply with ITAR regulations by obtaining the necessary licenses, maintaining detailed records, and submitting annual reports to the U.S. Department of State, Directorate of Defense Trade Controls.

The USML (United States Munitions List) classifies products and services that are subject to the International Traffic in Arms Regulations (ITAR). The USML is a comprehensive list of defense-related items, technologies, and services that are subject to ITAR export controls.

The USML classifies products and services into 21 categories, including:

- Firearms, Close Assault Weapons and Combat Shotguns
- Guns and Armament
- Ammunition/Ordnance
- Launch Vehicles, Guided Missiles, Ballistic Missiles, Rockets, Torpedoes, Bombs, and Mines
- Explosives and Energetic Materials, Propellants, Incendiary Agents, and Their Constituents
- Vessels of War and Special Naval Equipment
- Tanks and Military Vehicles
- Aerospace and Propulsion
- Electronic and Electro-Optical Systems, Including Guidance and Control Equipment
- Auxiliary Military Equipment
- Toxicological Agents, Including Chemical Agents, Biological Agents, and Associated Equipment
- Classification Proprietary to the United States Government
- Submersible Vessels, Oceanographic and Associated Equipment
- Materials and Miscellaneous Articles
- Directed Energy Weapons Systems
- Classified Articles, Technical Data and Defense Services Not Otherwise Enumerated
- Technical Data
- Defense Services
- Articles, Technical Data, and Defense Services for Protective Personnel
- Spacecraft Systems and Associated Equipment
- Systems, Equipment and Components, Test, Inspection and Production Equipment, and Related Software and Technology.

The categories are further broken down into specific items and services, along with the corresponding definitions and definitions of related terms. It is important to accurately determine the classification of a product or service to ensure compliance with ITAR regulations and to determine if a license or other authorization is required for export.

In review, ITAR regulations are a complex set of regulations that must be understood and complied with by all parties involved in the manufacture, export, import, transfer, or brokering of defense related items, services, and technology. Compliance managers in the defense industry must have a thorough understanding of ITAR regulations and their requirements in order to achieve ITAR compliance.

Definition of ITAR-Controlled Items, Services, and Technology

The International Traffic in Arms Regulations (ITAR) control the export of defense-related items, services, and technology. ITAR-controlled items, services, and technology are listed on the U.S. Munitions List (USML), which is located in 22 CFR Part 121 of the ITAR regulations.

ITAR-controlled items include a wide range of physical goods and technical data, including weapons systems, ammunition, defense articles, and related technical data. ITAR-controlled services include defense services related to the design, development, production, repair, overhaul, maintenance, modification, and testing of ITAR-controlled items. ITAR-controlled technology includes technical data related to ITAR-controlled items, as well as software and software source code related to the design, development, production, repair, overhaul, maintenance, modification, and testing of ITAR-controlled items.

ITAR licensing requirements are determined based on the nature of the item, service, or technology being exported, the end-use of the item, service, or technology, and the end-user of the item, service, or technology. ITAR licensing requirements may also be influenced by the presence of foreign nationals or foreign-origin items in the manufacturing or research and development process.

Use Cases:

A defense contractor is engaged in the manufacture of ITAR-controlled weapons systems. The defense contractor must comply with ITAR regulations by obtaining the necessary licenses, maintaining detailed records, and submitting annual reports to the U.S. Department of State, Directorate of Defense Trade Controls.

A defense contractor is providing ITAR-controlled defense services related to the design, development, and production of ITAR-controlled items. The defense contractor must comply with ITAR regulations by obtaining the necessary licenses, maintaining detailed records, and submitting annual reports to the U.S. Department of State, Directorate of Defense Trade Controls.

A defense contractor is exporting ITAR-controlled technical data related to the design, development, and production of ITAR-controlled items. The defense contractor must comply with ITAR regulations by obtaining the necessary licenses, maintaining detailed records, and submitting annual reports to the U.S. Department of State, Directorate of Defense Trade Controls.

In addition to the traditional ITAR-controlled items, services, and technology, it is also important to consider cyber security threats and the impact of ITAR regulations on the transfer of ITAR-controlled information through digital means.

As the use of digital communication and data storage has increased, the risk of unauthorized access to ITAR-controlled information has also increased.

ITAR regulations require that persons subject to ITAR comply with additional security measures, such as encryption, access controls, and network security, in order to protect ITAR-controlled information from unauthorized access and transfer. Compliance managers must ensure that their organizations have implemented adequate cyber security measures to comply with ITAR regulations and protect ITAR-controlled information from cyber security threats.

In review, ITAR-controlled items, services, and technology are an extensive and complex set of regulations that must be understood and complied with by all parties involved in the manufacture, export, import, transfer, or brokering of defense-related items, services, and technology. Compliance managers must have a thorough understanding of ITAR-controlled items, services, and technology and their requirements in order to achieve ITAR compliance.

ITAR Licensing Requirements

The International Traffic in Arms Regulations (ITAR) require licensing for the export of defense-related items, services, and technology. Licensing requirements are determined based on the nature of the item, service, or technology being exported, the end-use of the item, service, or technology, and the end-user of the item, service, or technology.

Compliance managers must have a thorough understanding of the ITAR licensing requirements in order to ensure compliance with ITAR regulations. Some common licensing requirements include:

End-Use Licensing Requirements: End-use licensing requirements specify the authorized end-uses of ITAR-controlled items, services, and technology. End-use licensing requirements may include restrictions on the use of ITAR-controlled items, services, and technology for military, nuclear, or chemical/biological purposes.

End-User Licensing Requirements: End-user licensing requirements specify the authorized end-users of ITAR-controlled items, services, and technology. End-user licensing requirements may include restrictions on the transfer of ITAR-controlled items, services, and technology to prohibited countries or to individuals and entities designated by the U.S. Department of Treasury's Office of Foreign Assets Control.

Technical Data Licensing Requirements: Technical data licensing requirements specify the conditions under which ITAR-controlled technical data may be exported. Technical data licensing requirements may include restrictions on the transfer of technical data related to the design, development, production, repair, overhaul, maintenance, modification, and testing of ITAR-controlled items.

Brokering Licensing Requirements: Brokering licensing requirements specify the conditions under which ITAR-controlled items, services, and technology may be brokered. Brokering licensing requirements may include restrictions on the transfer of ITAR-controlled items, services, and technology through intermediaries, including agents, brokers, and representatives.

Temporary Importation Licensing Requirements: Temporary importation licensing requirements specify the conditions under which ITAR-controlled items, services, and technology may be temporarily imported into the United States. Temporary importation licensing requirements may include restrictions on the re-export of ITAR-controlled items, services, and technology, as well as requirements for obtaining a Temporary Import License from the U.S. Department of State, Directorate of Defense Trade Controls.

In review, compliance managers must have a thorough understanding of ITAR licensing requirements in order to ensure compliance with ITAR regulations. ITAR licensing requirements are complex and may vary based on the specific circumstances of each export, including the nature of the item, service, or technology being exported, the end-use of the item, service, or technology, and the end-user of the item, service, or technology.

ITAR Recordkeeping and Reporting Requirements

The International Traffic in Arms Regulations (ITAR) require companies engaged in the export of defense-related items, services, and technology to maintain accurate records and to submit reports as necessary to the U.S. Department of State, Directorate of Defense Trade Controls. Recordkeeping and reporting requirements are an essential component of ITAR compliance and are critical to ensuring that the U.S. government is able to monitor the export of defense-related items, services, and technology.

Recordkeeping Requirements

ITAR-controlled items, services, and technology: Companies must maintain records of all ITAR-controlled items, services, and technology that are exported, including a description of the item, service, or technology, the name and address of the recipient, and the date and place of export.

Licenses and agreements: Companies must maintain records of all licenses and agreements related to ITAR-controlled items, services, and technology, including licenses and agreements issued by the U.S. Department of State, Directorate of Defense Trade Controls.

End-use and end-user information: Companies must maintain records of end-use and end-user information for ITAR-controlled items, services, and technology, including the authorized end-use and end-user of the item, service, or technology.

Technical data: Companies must maintain records of all technical data related to ITAR-controlled items, services, and technology, including data related to the design, development, production, repair, overhaul, maintenance, modification, and testing of ITAR-controlled items.

Brokering activities: Companies must maintain records of all brokering activities related to ITAR-controlled items, services, and technology, including records of the names and addresses of intermediaries, including agents, brokers, and representatives.

Reporting Requirements

Annual Report: Companies must submit an annual report to the U.S. Department of State, Directorate of Defense Trade Controls that provides information on all ITAR-controlled items, services, and technology that have been exported during the previous year. The annual report must include information on the nature of the item, service, or technology, the name and address of the recipient, and the date and place of export.

End-Use Reporting: Companies must report changes in the end-use or end-user of ITAR-controlled items, services, and technology to the U.S. Department of State, Directorate of Defense Trade Controls within 30 days of the change.

Brokering Reporting: Companies must report brokering activities related to ITAR-controlled items, services, and technology to the U.S. Department of

State, Directorate of Defense Trade Controls within 30 days of the brokering activity.

In review, ITAR recordkeeping and reporting requirements are an essential component of ITAR compliance. Companies engaged in the export of defense-related items, services, and technology must maintain accurate records and submit reports to the U.S. Department of State, Directorate of Defense Trade Controls as required. Compliance managers must have a thorough understanding of ITAR recordkeeping and reporting requirements in order to ensure compliance with ITAR regulations.

ITAR Compliance Obligations

The International Traffic in Arms Regulations (ITAR) establish a comprehensive regulatory framework for the export of defense-related items, services, and technology. To comply with ITAR regulations, companies must fulfill a range of obligations, including but not limited to:

Registration: Companies must register with the U.S. Department of State, Directorate of Defense Trade Controls (DDTC) in order to be eligible to engage in ITAR-controlled activities. The registration process involves the submission of a detailed application, along with the payment of an annual fee.

Licensing: Companies must obtain a license from DDTC prior to engaging in the export of ITAR-controlled items, services, or technology. The licensing process involves the submission of a detailed application, along with the payment of a processing fee. The license must be approved by DDTC before the export can take place.

Recordkeeping: Companies must keep detailed records of all ITAR-controlled exports and transfers, including the type and quantity of items or services exported, the recipient, the end-use, and the value of the export. The records must be kept for a minimum of five years from the date of the export or transfer.

Reporting: Companies must report any suspected or actual violations of ITAR regulations to DDTC within a specified timeframe. The report should include a detailed description of the suspected or actual violation, including the type and quantity of items or services involved, the recipient, and the end-use.

Security: Companies must implement appropriate security measures to protect ITAR-controlled items, services, and technology from unauthorized access, theft, or diversion. The security measures should be commensurate with the level of risk involved and should be reviewed and updated regularly.

Sanctions screening: Companies must screen all parties involved in ITAR-controlled activities, including employees, agents, partners, and customers, against the U.S. government's sanctions and export control lists. The screening should be conducted on a regular basis and should be documented in the company's records.

End-use monitoring: Companies must monitor the end-use of ITAR-controlled items, services, and technology to ensure that they are not used for unauthorized purposes. The monitoring should be conducted on a regular basis and should be documented in the company's records.

Technical assistance agreements: Companies must obtain approval from DDTC for any technical assistance agreements that involve ITAR-controlled items, services, or technology. The technical assistance agreement should be in writing and should be approved by DDTC before the technical assistance can be provided.

In review, compliance with ITAR regulations is a complex and ongoing process that requires a comprehensive understanding of the regulations and a commitment to compliance from all parties involved. Companies must fulfill a range of obligations, including registration, licensing, recordkeeping, reporting, security, sanctions screening, end use monitoring, and technical assistance agreements. Compliance managers must be knowledgeable and experienced in the compliance requirements and obligations of ITAR regulations in order to ensure compliance with the regulations.

Penalties for ITAR Non-Compliance

The International Traffic in Arms Regulations (ITAR) establish a comprehensive regulatory framework for the export of defense-related items, services, and technology. Compliance with ITAR regulations is mandatory and failure to comply with the regulations can result in serious consequences, including civil and criminal penalties.

Civil penalties: Companies that violate ITAR regulations may be subject to civil penalties, which can include fines of up to $1 million per violation, exclusion from future exports, and suspension or revocation of the company's ITAR registration. Civil penalties may also include the imposition of denial orders, which prohibit the company from participating in ITAR-controlled activities for a specified period of time.

A San Diego local, Joe Sery, admitted to his guilt in federal court on June 9, 2022 for his involvement in a conspiracy. The former CEO and owner of Tungsten Heavy Powder & Parts (THPP), Sery plead guilty to violating the United States' federal export laws through the International Traffic in Arms Regulations (ITAR) by exporting defense articles to countries such as the People's Republic of China and the Republic of India, without obtaining the necessary licenses from the U.S. Department of State. Sery, who was well-educated about the restrictions of U.S. export control laws, acknowledged in his plea agreement that he had knowingly exported items and data listed on the U.S. Munitions List, which are restricted from unlicensed exports.

"This arrest highlights the outstanding partnerships between HSI and the Department of Defense's investigative agencies who work tirelessly every day to ensure our protected military technology and weaponry are not used by foreign actors against our warfighters and allies on the battlefield," said HSI San Diego Special Agent in Charge Chad Plantz. "This arrest sends a clear message that those entrusted with our country's military technology and weaponry will be held responsible for its safeguarding."

Criminal penalties: Individuals who violate ITAR regulations may be subject to criminal penalties, which can include fines of up to $1 million per violation and imprisonment for up to 20 years. Criminal penalties may also include the imposition of debarment, which prohibits the individual from participating in ITAR-controlled activities for a specified period of time.

Reputational damage: Companies that violate ITAR regulations may also suffer significant reputational damage, which can impact their ability to do business in the future. Companies that are found to have violated ITAR regulations may be subject to negative publicity, loss of credibility, and loss of business opportunities.

Loss of clearance: Companies and individuals who violate ITAR regulations may also lose their security clearance, which can impact their ability to participate in government contracts and other sensitive projects.

Compliance costs: Companies that violate ITAR regulations may also face significant compliance costs, including the cost of implementing and maintaining an ITAR compliance program, the cost of defending against enforcement actions, and the cost of remediating any harm caused by the violation.

In review, non-compliance with ITAR regulations can result in serious consequences, including civil and criminal penalties, reputational damage, loss of clearance, and compliance costs. Compliance managers must be aware of the penalties for non-compliance and must take all necessary steps to ensure compliance with ITAR regulations in order to avoid these consequences.

Importance of ITAR Compliance for Compliance Managers

International Traffic in Arms Regulations (ITAR) regulate the export of defense-related items, services, and technology from the United States. As a compliance manager, it is critical to understand the importance of ITAR compliance, as non-compliance with ITAR regulations can result in serious consequences, including civil and criminal penalties, reputational damage, loss of clearance, and compliance costs.

Legal Obligations: Compliance with ITAR regulations is mandatory, and companies that violate ITAR regulations may be subject to civil and criminal penalties, including fines, imprisonment, exclusion from future exports, and suspension or revocation of the company's ITAR registration. Compliance managers must understand the legal obligations associated with ITAR compliance and take all necessary steps to ensure compliance.

Reputational Damage: Companies that violate ITAR regulations may suffer significant reputational damage, which can impact their ability to do business in the future. Companies that are found to have violated ITAR regulations may be subject to negative publicity, loss of credibility, and loss of business opportunities. Compliance managers must take steps to protect the company's reputation and minimize the risk of reputational damage.

Compliance Costs: Companies that violate ITAR regulations may face significant compliance costs, including the cost of implementing and maintaining an ITAR compliance program, the cost of defending against enforcement actions, and the cost of remediating any harm caused by the violation. Compliance managers must understand the costs associated with non-compliance and take steps to minimize these costs.

Cybersecurity Threats: ITAR-controlled items, services, and technology may also be targeted by foreign governments, criminal organizations, and individuals seeking to obtain sensitive information or technology. Compliance managers must be aware of the cybersecurity risks associated with ITAR-controlled items and take steps to protect against these risks.

Government Contracts: Companies and individuals who participate in government contracts may be subject to ITAR regulations, and non-compliance with ITAR regulations can result in the loss of clearance, suspension or revocation of the company's ITAR registration, and exclusion from future government contracts. Compliance managers must understand the importance of ITAR compliance for companies and individuals who participate in government contracts.

In review, ITAR compliance is critical for compliance managers, as non-compliance with ITAR regulations can result in serious consequences, including civil and criminal penalties, reputational damage, loss of clearance, and compliance costs. Compliance managers must understand the importance of ITAR compliance and take all necessary steps to ensure compliance.

Consequences of ITAR Non-Compliance

ITAR non-compliance can result in serious consequences for both individuals and companies, including large fines, penalties, and even imprisonment. These consequences are designed to be severe in order to deter individuals and companies from violating ITAR regulations, which are put in place to protect national security interests.

Fines: One of the most significant consequences of ITAR non-compliance is financial penalties. Companies and individuals that are found to have violated ITAR regulations may face fines of up to $1,000,000 per violation, with a maximum of $100,000,000 per incident. Fines can also be assessed on a per item basis, which can quickly add up for companies with large numbers of ITAR-controlled items.

Penalties: In addition to fines, individuals and companies may face other penalties, such as the denial of export privileges, the denial of government contracts, and even imprisonment. These penalties can be devastating for individuals and companies that rely on exports or government contracts for their livelihoods.

Imprisonment: In extreme cases, individuals who violate ITAR regulations may face imprisonment. This is typically reserved for individuals who have knowingly and willfully violated ITAR regulations, or who have attempted to export ITAR-controlled items to prohibited countries or persons.

Damage to Reputation: ITAR non-compliance can also result in damage to a company's reputation. This can make it difficult for the company to obtain new contracts or to continue doing business with existing customers. Companies may also face difficulties in attracting and retaining employees, as well as increased regulatory scrutiny.

Loss of Export Privileges: Companies that violate ITAR regulations may face the loss of their export privileges, which can have a significant impact on their ability to do business. This can result in a loss of revenue, as well as the need to find new markets for their products.

It is important to note that ITAR non-compliance can result in serious consequences, regardless of whether the violation was intentional or accidental. It is the responsibility of compliance managers to ensure that their companies are in compliance with ITAR regulations, and to take the necessary steps to prevent ITAR violations from occurring.

Developing an ITAR Compliance Program

The International Traffic in Arms Regulations (ITAR) require companies engaged in the export of defense-related items, services, and technology to comply with a range of regulatory requirements. To ensure compliance with ITAR regulations, companies must develop and implement an effective ITAR compliance program.

An ITAR compliance program should include the following key elements:

Management commitment: A commitment from senior management is essential for the success of an ITAR compliance program. The compliance program should be integrated into the company's overall business strategy and operations, and senior management should provide the necessary resources and support for the program to be successful.

Employee training: All employees should receive training on ITAR regulations and the company's ITAR compliance program. The training should be comprehensive, covering the regulations and their application to the company's operations, as well as the procedures for reporting suspected or actual violations.

Documentation: The company should establish written policies and procedures that outline its ITAR compliance program. The documentation should include procedures for licensing, recordkeeping, and reporting, as well as procedures for handling suspected or actual violations.

Designation of a compliance officer: The company should designate a compliance officer who is responsible for overseeing the implementation and maintenance of the ITAR compliance program. The compliance officer should have the necessary authority, resources, and training to perform their duties effectively.

Regular audits and self-assessments: The company should conduct regular internal audits and self-assessments to evaluate the effectiveness of its ITAR compliance program. The results of these audits and self-assessments should be used to make improvements to the program as necessary.

Sanctions for violations: The company should establish a system of sanctions for employees who violate ITAR regulations or the company's ITAR compliance program. The sanctions should be commensurate with the seriousness of the violation and should serve as a deterrent to future violations.

Continuous improvement: The company should continuously improve its ITAR compliance program by regularly reviewing and updating its policies and procedures, conducting internal audits and self-assessments, and incorporating best practices from other companies.

In review, an effective ITAR compliance program is essential for companies engaged in the export of defense related items, services, and technology. The compliance program should be comprehensive, covering all relevant regulations and requirements, and should be integrated into the company's overall business strategy and operations. Compliance managers must be knowledgeable and experienced in the development and implementation of ITAR compliance programs in order to ensure compliance with ITAR regulations.

Use Cases and Best Practices for ITAR Compliance

In this chapter, we will discuss real-life scenarios and best practices for ITAR compliance. These use cases are designed to provide practical examples of how the International Traffic in Arms Regulations (ITAR) are applied in the real world and to provide guidance on how to maintain compliance in a variety of situations. *Due to complexity of ITAR use cases, please contact Cleared Systems for details about each use case.*

Use Case 1: Exporting ITAR-Controlled Items to Foreign Customers

In this use case, we will examine the process of exporting ITAR-controlled items to foreign customers. We will discuss the licensing requirements, recordkeeping obligations, and reporting requirements that must be met in order to comply with ITAR regulations.

Use Case 2: Providing ITAR-Controlled Services to Foreign Customers

In this use case, we will examine the process of providing ITAR-controlled services to foreign customers. We will discuss the licensing requirements, recordkeeping obligations, and reporting requirements that must be met in order to comply with ITAR regulations.

Use Case 3: Sharing ITAR-Controlled Technology with Foreign Persons

In this use case, we will examine the process of sharing ITAR-controlled technology with foreign persons. We will discuss the licensing requirements, recordkeeping obligations, and reporting requirements that must be met in order to comply with ITAR regulations.

Use Case 4: Maintaining ITAR Compliance during Mergers and Acquisitions

In this use case, we will examine the process of maintaining ITAR compliance during mergers and acquisitions. We will discuss the importance of due diligence and how to ensure that the acquired company is in compliance with ITAR regulations.

Use Case 5: Conducting ITAR-Related Business with Foreign Governments

In this use case, we will examine the process of conducting ITAR-related business with foreign governments. We will discuss the licensing requirements, recordkeeping obligations, and reporting requirements that must be met in order to comply with ITAR regulations.

Use Case 6: Ensuring ITAR Compliance during Research and Development Projects

In this use case, we will examine the process of ensuring ITAR compliance during research and development projects. We will discuss the importance of developing clear protocols and procedures for ITAR compliance and how to maintain ITAR compliance throughout the R&D process.

Use Case 7: Maintaining ITAR Compliance during Supply Chain Management

In this use case, we will examine the process of maintaining ITAR compliance during supply chain management. We will discuss the importance of due diligence and how to ensure that suppliers are in compliance with ITAR regulations.

Use Case 8: Responding to ITAR-Related Government Investigations

In this use case, we will examine the process of responding to ITAR-related government investigations. We will discuss the importance of having an ITAR compliance program in place and how to respond to government inquiries and investigations in a compliant manner.

Use Case 9: Implementing ITAR Compliance Programs in Global Organizations

In this use case, we will examine the process of implementing ITAR compliance programs in global organizations. We will discuss the importance of developing global protocols and procedures for ITAR compliance and how to ensure that all operations, regardless of location, are in compliance with ITAR regulations.

> *In review, these use cases and best practices provide a comprehensive overview of how to maintain ITAR compliance in a variety of situations. By using these examples compliance managers can better understand how to apply the International Traffic in Arms Regulations (ITAR) in their organizations.*

Understanding Export Controls

Overview of Export Control Regulations

Export control regulations are laws and regulations that govern the export of certain goods, services, and technologies from one country to another. These regulations are designed to ensure that sensitive or strategically important items do not fall into the wrong hands and are not used in ways that could harm national security or foreign policy interests.

There are several different export control regimes in place around the world, each with its own unique set of regulations and requirements. Some of the most well-known export control regimes include the U.S. International Traffic in Arms Regulations (ITAR), the U.S. Export Administration Regulations (EAR), and the U.K. Strategic Export Controls.

Each export control regime is tailored to address specific security and foreign policy concerns and to regulate different types of goods, services, and technologies. For example, ITAR regulations focus primarily on the export of military and dual-use items, while EAR regulations are more focused on commercial and civilian items with potential military applications.

Compliance with export control regulations is a complex and constantly evolving task that requires a deep understanding of the regulations and the items being exported. In order to ensure compliance with these regulations, organizations must implement robust compliance programs that are designed to manage the risks associated with exporting.

Export control regulations are subject to change and updates, and it is important for organizations to stay up-to-date with the latest developments. Some of the key references for information on export control regulations include the relevant government agencies responsible for enforcing the regulations, such as the U.S. Department of Commerce and the U.S. Department of State. Industry associations, trade organizations, and publications also provide useful information and resources on export control regulations.

Key Components of Export Control Regulations

Export control regulations are a critical component of international trade, as they are designed to protect national security, promote foreign policy objectives, and prevent the spread of weapons of mass destruction. These regulations are used by governments around the world to control the export, re-export, and transfer of goods, technology, and services. In this chapter, we will provide an overview of the key components of export control regulations.

Proscribed Countries and Entities: One of the key components of export control regulations is the list of proscribed countries and entities. These lists are used by governments to restrict trade with countries and entities that are deemed to pose a threat to national security or foreign policy objectives. Examples of proscribed countries and entities include Iran, North Korea, and organizations that are designated as state sponsors of terrorism.

Dual-Use Goods, Technology, and Services: Dual-use goods, technology, and services are those that have both commercial and military applications. These items are subject to export control regulations, as they have the potential to be used in the development of weapons of mass destruction. Examples of dual-use items include software, hardware, and technology that is related to encryption, missile guidance systems, and nuclear technology.

End-Use Restrictions: End-use restrictions are a critical component of export control regulations, as they limit the use of goods, technology, and services to specific end-uses. These restrictions are designed to prevent the unauthorized use of items for military or other prohibited purposes.

Licensing Requirements: Licensing requirements are another key component of export control regulations, as they ensure that the export, re-export, and transfer of goods, technology, and services are authorized and approved by the appropriate government agency. These licensing requirements vary by country, but typically involve the submission of a license application and a review of the proposed transaction.

Compliance Obligations: Compliance obligations are a critical component of export control regulations, as they require companies to ensure that they are adhering to all applicable regulations and requirements. These obligations include recordkeeping, reporting, and self-assessment requirements.

Penalties for Non-Compliance: Penalties for non-compliance with export control regulations can be severe, and may include fines, imprisonment, and debarment from participating in international trade.

In review, export control regulations are an essential component of international trade, as they protect national security and promote foreign policy objectives. It is important for compliance managers to be familiar with the key components of these regulations, as well as their obligations and responsibilities under the regulations.

Export Control Compliance Obligations

Export control regulations aim to prevent the export of goods, services, and technologies that could be used for military purposes or otherwise harm national security. Compliance with these regulations is a critical obligation for companies and organizations that deal with controlled items, services, and technology. In this chapter, we will discuss the key components of export control compliance obligations and how they are enforced.

Overview of Export Control Compliance Obligations

Export control compliance obligations are the set of rules and regulations that companies and organizations must follow in order to ensure that their exports of controlled items, services, and technology are in line with national security and foreign policy objectives. These obligations are established by government agencies such as the Department of Commerce, Department of State, and Department of Treasury.

Understanding Export Licensing Requirements

One of the key components of export control compliance obligations is understanding the licensing requirements for controlled items, services, and technology. Companies and organizations must obtain the appropriate licenses from the government agencies responsible for enforcing export controls before they can export any controlled items, services, or technology.

Export Recordkeeping and Reporting Requirements

In addition to licensing requirements, companies and organizations must also comply with recordkeeping and reporting requirements for their exports. This includes maintaining records of all exports, including the type of item exported, the destination country, the value of the export, and the name of the recipient.

Companies and organizations must also report any changes to their export activities to the relevant government agencies.

Developing an Export Control Compliance Program

To ensure compliance with export control regulations, companies and organizations must develop and implement an effective export control compliance program. This program should include procedures for obtaining licenses, maintaining records, reporting changes in export activities, and conducting due diligence on potential business partners. Companies and organizations should also provide training to employees on export control compliance obligations to ensure that they are aware of the regulations and their responsibilities.

Importance of Export Control Compliance

Export control compliance is not only a legal obligation, but it is also essential to protect national security and foreign policy objectives. Companies and organizations that fail to comply with export control regulations can face severe penalties, including fines, imprisonment, and loss of business opportunities. Therefore, it is important for companies and organizations to take export control compliance obligations seriously and implement an effective compliance program to avoid potential penalties.

Navigating the Complexities of ITAR and Export Controls

Identifying ITAR-Controlled Items and Services

The International Traffic in Arms Regulations (ITAR) are a set of regulations that govern the export of defense-related articles, services, and technologies from the United States. As a compliance manager, it is essential to understand what is considered an ITAR-controlled item or service, as these are subject to the strict export control requirements outlined in the regulations.

In general, ITAR-controlled items include defense articles and services, such as military equipment, firearms, ammunition, and related technical data. This also includes any item or service that has been specifically designed or modified for military applications. ITAR-controlled items are listed in the U.S. Munitions List (USML), which is a comprehensive list of items that are subject to ITAR regulations.

It is important to note that ITAR regulations apply to the export of not just physical items, but also technical data and services that are related to the production or maintenance of defense articles. This includes everything from technical specifications and blueprints, to training and engineering services.

To identify ITAR-controlled items and services, it is recommended that compliance managers conduct a thorough review of their products, services, and operations. This can include reviewing the USML and consulting with experts in the field to ensure that all ITAR-related obligations are properly understood and met.

In addition, compliance managers should also establish internal controls and procedures for identifying and tracking ITAR-controlled items and services. This can include implementing internal audits, training programs for employees, and developing clear processes for documenting and reporting ITAR-related activities.

By taking these steps, compliance managers can ensure that their organization is in compliance with ITAR regulations and that all exports of ITAR-controlled items and services are properly authorized and regulated.

Determining ITAR Licensing Requirements

In order to ensure compliance with International Traffic in Arms Regulations (ITAR), it is important to understand the licensing requirements for exporting ITAR-controlled items and services. The ITAR requires that companies obtain a license or other written authorization from the U.S. Department of State before exporting any ITAR-controlled item or service.

There are a number of factors that go into determining whether a license is required for a particular export, including the type of item or service being exported, the end-use of the item or service, the end-user of the item or service, and the destination country. In some cases, a license may not be required, but a written exemption or other authorization may be required.

It is the responsibility of the compliance manager to understand the licensing requirements for the specific items and services that their company exports, and to ensure that the appropriate licenses or authorizations are obtained before any exports take place. To determine the licensing requirements for your company, you should consult the ITAR and other relevant regulations, as well as seek guidance from experienced export control professionals.

It is also important to keep in mind that the licensing requirements can change over time, so it is essential to stay up-to-date on the latest regulations and requirements. Regular training and education on export control regulations can help ensure that you and your team are aware of any changes and can continue to operate in compliance with ITAR regulations.

Evaluating Export Control Risks

One of the key responsibilities of compliance managers is to assess and manage risks associated with the export of ITAR-controlled items and services. This requires a thorough understanding of the international trade regulations and a proactive approach to mitigating risks.

In this chapter, we will discuss the process of evaluating export control risks and provide practical tips for compliance managers.

First, we will start by identifying the different types of export control risks, including legal, reputational, operational, and commercial risks. Each type of risk presents unique challenges and requires a different approach to mitigation.

Next, we will discuss how to perform a risk assessment to identify the likelihood and impact of potential risks. This includes analyzing internal and external factors, such as company policies and procedures, customer relationships, and supply chain operations.

Performing a risk assessment is an important step in ensuring ITAR and Export Control Compliance. Here are the steps to perform a risk assessment:

1. **Identify your products and services:** Determine which of your products and services are subject to ITAR and Export Control regulations by reviewing the USML.
2. **Evaluate your business operations:** Consider all aspects of your business operations, including procurement, manufacturing, sales, marketing, research and development, and logistics.
3. **Identify potential risks:** Evaluate each aspect of your business operations to identify potential risks, such as the transfer of controlled technology to foreign nationals, unauthorized exports of controlled items, or inadequate recordkeeping and reporting.
4. **Assess the likelihood and impact of each risk:** Consider the likelihood of each risk occurring and the potential impact of the risk on your business, including financial and reputational consequences.
5. **Evaluate and prioritize risks:** Based on the likelihood and impact of each risk, prioritize the risks that need to be addressed first.
6. **Develop a risk mitigation plan:** For each prioritized risk, develop a plan to mitigate the risk. This may involve implementing new policies and procedures, performing additional training, or engaging third-party consultants.
7. **Monitor and evaluate the effectiveness of your risk mitigation plan:** Regularly monitor and evaluate the effectiveness of your risk mitigation plan to ensure that it remains relevant and effective in addressing the risks faced by your business.
8. **Revise the risk assessment as necessary:** As your business operations evolve, regularly revisit your risk assessment to ensure that it remains accurate and up-to-date.

By performing a comprehensive risk assessment, you can identify potential risks and develop a plan to mitigate those risks, helping to ensure that your business is fully compliant with ITAR and Export Control regulations.

Once the risks have been identified, compliance managers can implement a risk mitigation plan that addresses each risk and implements controls to minimize or eliminate the impact of the risks.

Finally, we will discuss the importance of monitoring and reviewing the effectiveness of the risk mitigation plan on an ongoing basis, and updating the plan as needed to address new or changing risks.

Overall, evaluating export control risks is a critical component of an ITAR compliance program, and requires a proactive and continuous approach to mitigate risks and maintain compliance with ITAR regulations.

Developing ITAR and Export Control Compliance Strategies

As a compliance manager, it is important to understand the regulations that govern the export of defense-related items and technology, including the International Traffic in Arms Regulations (ITAR) and export control regulations. To ensure compliance, it is necessary to develop effective strategies to manage these regulations. This chapter will provide an overview of the key elements of ITAR and export control compliance strategies.

ITAR and export control compliance strategies should be based on a thorough understanding of the regulations and the items, services, and technology that are subject to control. This includes an evaluation of the specific risks and requirements associated with each item, service, or technology, and an assessment of the associated compliance obligations.

A key component of ITAR and export control compliance strategies is the development of effective policies and procedures. These should be designed to ensure that employees are aware of their obligations under the regulations and to minimize the risk of non-compliance. This may include policies and procedures related to licensing, recordkeeping, reporting, and the training of employees.

Another important aspect of ITAR and export control compliance strategies is the development of risk mitigation measures. This may include measures to ensure that only authorized employees have access to ITAR-controlled items, services, and technology, and to prevent unauthorized exports or re-exports. It may also include measures to ensure that adequate controls are in place to monitor exports, and to detect and report any potential violations of the regulations.

In addition, ITAR and export control compliance strategies should include ongoing monitoring and auditing processes to ensure ongoing compliance. This may include regular audits of policies, procedures, and records, and regular training sessions for employees.

Ultimately, effective ITAR and export control compliance strategies require a commitment to continuous improvement. This may include regularly reviewing and updating policies and procedures and conducting regular assessments of compliance risks. By taking a proactive approach to compliance, organizations can minimize the risk of non-compliance and ensure that they are in compliance with the regulations that govern the export of defense-related items and technology.

Best Practices for ITAR and Export Control Compliance

Employee Training and Awareness

A crucial component of a successful ITAR and export control compliance program is the education and training of employees. Employees at all levels, from executives to administrative staff, need to understand their role in complying with ITAR and export control regulations. This chapter will explore the importance of employee training and awareness and provide best practices for conducting effective training programs.

The purpose of employee training is to ensure that all employees are knowledgeable about ITAR and export control regulations and understand their obligations under these regulations. It is also important to educate employees on the potential consequences of non-compliance, including civil and criminal penalties, damage to a company's reputation, and loss of business opportunities.

When developing a training program, it is important to consider the specific needs of your company and the types of ITAR-controlled items and services your company deals with. A one-size-fits-all approach may not be appropriate for all companies, and customization may be necessary to ensure that employees receive relevant and practical training.

The following are best practices for employee training and awareness programs:

- Develop clear policies and procedures that define employee obligations under ITAR and export control regulations.
- Provide regular training sessions to employees, and make sure that training is updated to reflect any changes in regulations or company policies.
- Ensure that all employees, including new hires and temporary employees, receive ITAR and export control training as part of their onboarding process.
- Provide training materials, such as handbooks, presentations, and videos, that employees can refer to as needed.
- Assign a designated compliance officer who is responsible for conducting training and monitoring employee compliance with ITAR and export control regulations.
- Consider conducting regular audits and assessments to measure the effectiveness of your training program and identify areas for improvement.

By conducting effective employee training and awareness programs, companies can ensure that their employees are knowledgeable about ITAR and export control regulations and understand their role in complying with these regulations. This will help companies avoid potential penalties for non-compliance and protect their reputation.

Developing Internal Controls and Processes

Internal controls and processes are an essential part of any ITAR and Export Control Compliance Program. These controls and processes provide a framework for ensuring that the program is implemented effectively and consistently across the organization. This chapter will cover key considerations for developing internal controls and processes and provide practical examples of effective internal controls and processes.

Assessing your organization's risk profile: Before developing internal controls and processes, it is important to understand the specific risks and vulnerabilities associated with your organization's activities. This will help you to identify areas where internal controls and processes are needed to mitigate those risks.

Defining roles and responsibilities: To ensure effective implementation of internal controls and processes, it is important to clearly define the roles and responsibilities of all parties involved. This includes the compliance manager, department heads, and all employees who are involved in the handling of ITAR-controlled items and services.

Establishing documentation requirements: Effective documentation is key to demonstrating compliance with ITAR and Export Control regulations. This includes documentation of the procedures, policies, and processes in place, as well as any relevant agreements, licenses, and export declarations.

Implementing a system for ongoing monitoring and reporting: To ensure ongoing compliance with ITAR and Export Control regulations, it is important to have a system in place for monitoring and reporting. This includes regular reviews of internal controls and processes to ensure that they are effective and regularly updated, as well as regular reporting of any non-compliant activities to the relevant authorities.

Providing ongoing training and support: Finally, it is important to provide ongoing training and support to all employees involved in the handling of ITAR-controlled items and services. This includes regular training sessions to ensure that all employees understand the requirements of ITAR and Export Control regulations, as well as ongoing support to help employees navigate the complex requirements of these regulations.

By developing effective internal controls and processes, compliance managers can ensure that their ITAR and Export Control Compliance Program is implemented effectively and consistently across the organization. This will help to minimize the risk of non-compliance and ensure that the organization is able to meet its obligations under ITAR and Export Control regulations.

Maintaining Proper Records and Documentation

Proper recordkeeping and documentation are essential components of ITAR and export control compliance. Compliance managers must ensure that all relevant information is accurately recorded and easily accessible to avoid any potential violations. In this chapter, we will examine the importance of recordkeeping and documentation in ITAR and export control compliance, as well as provide guidance on what information should be recorded and how it should be stored.

The ITAR regulations require that companies maintain records of all defense articles and services exported, including the name of the recipient, the date of shipment, and the purpose of the export. This information must be kept for a minimum of five years from the date of export. In addition, companies must keep records of all license applications, approved licenses, and any other ITAR-related documents for at least five years from the date of expiration or revocation of the license.

Export control regulations also have similar recordkeeping requirements. For example, the Export Administration Regulations (EAR) require that companies keep records of all exports and reexports of controlled items, including the name and address of the recipient, the date of shipment, and the license or exemption used for the export.

To maintain proper records and documentation, companies should establish a system for tracking and storing all ITAR- and export control-related information. This may include an electronic database, physical file storage, or a combination of both. Companies should also ensure that all employees who are involved in the export process are trained on the recordkeeping requirements and understand the importance of accurate and timely recordkeeping.

In addition to recordkeeping, companies must also maintain proper documentation to support ITAR and export control compliance. This may include licenses, agreements, invoices, and other documents related to the export. Companies should ensure that all relevant documentation is kept in a secure and easily accessible location, and that all employees who may need access to the documentation are trained on the proper procedures for accessing and storing it.

In review, proper recordkeeping and documentation are critical components of ITAR and export control compliance. By establishing a system for tracking and storing information, maintaining accurate and timely records, and keeping proper documentation, compliance managers can ensure that their company remains compliant with all relevant regulations.

Working with Business Partners and Customers

When a company is involved in the international trade of defense articles and services, it is important to ensure that its business partners and customers are also in compliance with ITAR regulations. This is because non-compliance by these third parties can result in violations of ITAR regulations and lead to penalties for the company. In this chapter, we will discuss the best practices for working with business partners and customers to ensure ITAR compliance.

Conduct Due Diligence on Business Partners and Customers: Companies should conduct proper due diligence on their business partners and customers to ensure that they have a good understanding of their operations, products and services, and compliance with ITAR regulations. This can include reviewing their licenses, permits, and other certifications, as well as reviewing their compliance history.

Establishing ITAR Compliance Agreements: Companies can establish ITAR compliance agreements with their business partners and customers, which outline the responsibilities and obligations of both parties in ensuring ITAR compliance. These agreements should be clear and specific and should include provisions for regular compliance audits and reporting requirements.

Encouraging Compliance: Companies should encourage their business partners and customers to comply with ITAR regulations by providing training and education, and by including compliance requirements in their contracts and agreements.

Monitoring Compliance: Companies should monitor the compliance of their business partners and customers through regular audits, reviews, and reporting requirements. This helps to ensure that ITAR regulations are being followed and that the company remains in compliance.

Addressing Non-Compliance: In the event that a business partner or customer is found to be non-compliant with ITAR regulations, companies should have processes in place for addressing this non-compliance, including reporting it to the appropriate authorities and taking any necessary corrective actions.

In review, working with business partners and customers is a critical component of ITAR and export control compliance. Companies should take the necessary steps to ensure that their business partners and customers are in compliance, and that they have the proper processes and systems in place to monitor and address any non-compliance.

ITAR and Security

Importance of Cybersecurity

In the modern age, organizations have a vast amount of data stored in digital form that must be protected from unauthorized access and potential breaches. This is especially true for organizations that deal with sensitive information that is regulated by the International Traffic in Arms Regulations (ITAR). One of the critical challenges organizations face is ensuring the security of ITAR data when it comes to cybersecurity.

Microsoft GCC High: A Solution for ITAR Data Protection

One of the ways to protect ITAR data is to utilize Microsoft GCC High, which is a cloud-based solution provided by Microsoft. It is specifically designed to meet the regulatory and security requirements of government organizations and is an essential tool in safeguarding ITAR data. Microsoft GCC High provides a secure environment for data storage, handling, and processing. The platform is compliant with various international standards, including ITAR, and is equipped with state-of-the-art security features that protect data from cyber-attacks, unauthorized access, and data breaches.

Key Features of Microsoft GCC High

Microsoft GCC High is a highly secure and reliable platform that provides a wide range of features to protect ITAR data. Some of the key features include:

- **Advanced encryption:** Microsoft GCC High utilizes advanced encryption technologies to secure data in transit and at rest. Data is encrypted using the latest encryption algorithms and standards to ensure that data remains confidential and protected from potential cyber-attacks.

- **Data loss prevention:** Microsoft GCC High provides data loss prevention (DLP) capabilities that prevent sensitive information from being accidentally leaked. This helps to ensure that data is only accessible by authorized users and that sensitive information remains protected.
- **Multifactor authentication:** Microsoft GCC High uses multifactor authentication to secure access to data. Users are required to provide multiple forms of authentication, such as a password and a security token, to access ITAR data. This provides an extra layer of security and helps to prevent unauthorized access.
- **Continuous monitoring and auditing:** Microsoft GCC High is equipped with continuous monitoring and auditing capabilities that provide real-time visibility into data access and usage. This helps to identify potential security threats and take action to prevent potential data breaches.

Benefits of Microsoft GCC High

Microsoft GCC High provides a wide range of benefits to organizations that need to protect ITAR data. Some of the key benefits include:

- **Compliance:** Microsoft GCC High is ITAR compliant and helps organizations to comply with the regulatory requirements of the ITAR.
- **Security:** Microsoft GCC High provides a highly secure environment for ITAR data and helps to protect sensitive information from potential cyber-attacks and data breaches.
- **Accessibility:** Microsoft GCC High provides users with secure and convenient access to ITAR data from anywhere and at any time. This makes it easier for organizations to collaborate and work more efficiently.
- **Scalability:** Microsoft GCC High is a scalable solution that can adapt to the changing needs of organizations. This makes it an ideal solution for

organizations that need to store and manage increasing amounts of ITAR data.

In review, Microsoft GCC High is a highly secure and reliable solution that provides organizations with the ability to protect ITAR data from potential cyber-attacks and data breaches. With its advanced security features and compliance with ITAR, Microsoft GCC High is an essential tool for organizations that deal with sensitive information regulated by the ITAR.

Storing ITAR Data and Data Security

Although this book is mainly about the ITAR process, it is also important to discuss the importance of ITAR cybersecurity and data storage. The security and proper storage of ITAR-controlled data is of utmost importance in ensuring compliance with export control regulations. A data breach or unauthorized release of sensitive information can have severe consequences, including fines, loss of license, and damage to reputation. In this chapter, we will explore best practices for storing ITAR data and securing it against unauthorized access.

1. Identify ITAR-controlled data The first step in securing ITAR-controlled data is to identify what information is subject to the regulations. This information may include technical data, software, and other proprietary or confidential information related to defense articles, services, and technologies.
2. Store ITAR data securely Once you have identified ITAR-controlled data, it is essential to store it in a secure manner. This can include using secure file sharing platforms, encrypted hard drives, and firewalls. Additionally, you should limit access to ITAR data to only those employees who need it to perform their job functions.

3. Develop data security policies Your company should develop comprehensive data security policies that outline how ITAR data should be stored and protected. These policies should be reviewed regularly and updated as necessary to reflect changes in technology and the evolving threat landscape.
4. Use multi-factor authentication Multi-factor authentication (MFA) is a critical component of any data security program. MFA requires users to provide multiple forms of authentication, such as a password and a security token, to access ITAR data. This added layer of security helps to prevent unauthorized access to sensitive information.
5. Perform regular data security assessments Regular data security assessments can help to identify vulnerabilities and potential threats to ITAR data. These assessments should include internal audits, penetration testing, and vulnerability scans.
6. Encrypt ITAR data Encrypting ITAR data can provide an additional layer of protection against unauthorized access or theft. This can be done using encryption software, which encodes data in such a way that it can only be accessed with a decryption key.
7. Train employees on data security Employees play a critical role in ensuring the security of ITAR data. It is important to provide regular training and awareness programs to help employees understand the importance of data security and the role they play in protecting ITAR data.
8. Implement access controls Implementing access controls can help to prevent unauthorized access to ITAR data. This can include setting up user accounts, defining user roles, and monitoring access logs.

In review, storing ITAR data and securing it against unauthorized access is a critical aspect of export control compliance. By following these best practices, companies can help to ensure the confidentiality and security of sensitive information and maintain compliance with ITAR regulations.

Visitor Logs: Maintaining Compliance in ITAR and Export Control Regulations

Visitor logs are an important aspect of ITAR and Export Control compliance. Visitors to your facilities, such as contractors, business partners, or customers, may have access to ITAR-controlled information or materials. Keeping accurate and detailed visitor logs helps ensure that ITAR and Export Control regulations are being followed and that access to sensitive information is properly monitored and controlled.

Purpose of Visitor Logs

Visitor logs serve several important purposes in ITAR and Export Control compliance. First, they provide a record of who has had access to your facilities and ITAR-controlled information. This helps to track and monitor access to sensitive information and can be used to conduct audits or investigations if necessary.

Second, visitor logs provide a record of who has been trained on ITAR and Export Control regulations. By tracking who has received training, organizations can ensure that only authorized individuals have access to ITAR-controlled information.

Finally, visitor logs help to ensure that proper security protocols are being followed. This can include the use of security badges, signing non-disclosure agreements, or following specific procedures for accessing sensitive areas of your facilities.

Components of a Visitor Log

Visitor logs should include a variety of information to ensure that they are complete and useful for compliance purposes. Some key components to include in a visitor log are:

- Visitor's name and contact information
- Date and time of visit
- Purpose of visit
- Host's name and contact information
- Training received on ITAR and Export Control regulations
- Security protocols followed during the visit
- Signature of visitor and host

It is also important to note that visitor logs should be kept in a secure location and access to them should be restricted to authorized individuals.

Implementing a Visitor Log System

To effectively implement a visitor log system, organizations should follow these steps:

1. Determine the scope of your visitor log system. Will it cover all visitors to your facilities, or only those who will have access to ITAR-controlled information or materials?
2. Develop a standard form for visitor logs that includes the key components outlined above.
3. Train all personnel on the importance of visitor logs and how to properly complete them.
4. Establish a process for logging visitors in and out of your facilities, and make sure that all visitors are aware of this process.

5. Regularly review and update your visitor log system to ensure that it remains effective and efficient.

In review, visitor logs are an important aspect of ITAR and Export Control compliance. By keeping accurate and detailed visitor logs, organizations can ensure that access to sensitive information is properly monitored and controlled, and that proper security protocols are being followed. By following the steps outlined above, organizations can effectively implement a visitor log system that helps them maintain compliance with ITAR and Export Control regulations.

Conclusion

Summary of Key Points

The "ITAR and Export Controls Fundamentals: A Guide for Compliance Managers" book is a comprehensive guide for those who are responsible for ensuring compliance with international traffic in arms regulations (ITAR) and export control regulations. The book provides an overview of ITAR regulations and key components of export control regulations, followed by a discussion of the compliance obligations and best practices for ITAR and export control compliance.

The following key points were covered:

1. Understanding the International Traffic in Arms Regulations (ITAR) and the role of the Directorate of Defense Trade Controls (DDTC)
2. Determining ITAR-controlled items and services
3. Evaluating export control risks
4. Developing ITAR and export control compliance strategies
5. Best practices for ITAR and export control compliance
6. Employee training and awareness
7. Developing internal controls and processes
8. Maintaining proper records and documentation
9. Working with business partners and customers
10. Storing ITAR data and data security
11. Visitor logs
12. The importance of ITAR and cybersecurity

By following the guidelines provided in this book, organizations can ensure that they are compliant with ITAR and other US export control laws, thereby avoiding the legal and financial consequences of non-compliance. This guide has provided a solid foundation for compliance managers and other stakeholders to build on and ensure that they are fully equipped to navigate the complexities of ITAR and US export controls.

Appendix

1. ITAR (International Traffic in Arms Regulations): A set of regulations governing the export of defense-related articles, services, and technologies from the United States.
2. DDTC (Directorate of Defense Trade Controls): The government agency responsible for administering and enforcing ITAR.
3. USML (United States Munitions List): A list of defense-related articles, services, and technologies that are subject to ITAR and require an export license from the US State Department.
4. Export Control Compliance Program (ECP): A documented program that outlines a company's policies, procedures, and practices for ensuring compliance with ITAR and export control regulations.
5. End-User: The final recipient of a product or service, who will use or benefit from it.
6. Defense Article: Any item specified on the USML, including technical data, software, components, and finished products.
7. Defense Service: Any service, such as repair and maintenance, that involves the use of defense articles or technical data.
8. Export License: A government-issued permit allowing the export of defense-related articles, services, and technologies from the United States.
9. Risk Assessment: The process of evaluating and assessing the risks associated with exporting defense-related items and services.
10. Encryption: The process of encoding data so that it can only be deciphered by someone with the correct decryption key.

11. Firewall: A software or hardware system designed to protect a computer network from unauthorized access.
12. Visitor Log: A record of visitors to a facility, including the date and time of their visit, the purpose of their visit, and the person they met with.
13. Dual-Use Items: Items that have both civilian and military applications and may be subject to export controls.
14. Re-Export: The export of an item that has been imported into a country from another country, or the export of a product that contains components that originated from another country.
15. Transfer: The provision of a defense article or service to a foreign person within the United States, including by means of remote or electronic transmission.

ITAR and Export Control Compliance Resources

i. **ITAR Part 120 – Purpose and Definitions**
ii. **ITAR Part 121 – The United States Munitions List**
iii. **ITAR Part 122 – Registration of Manufacturers and Exporters**
iv. **ITAR Part 123 – Licenses for the Export of Defense Articles**
v. **ITAR Part 124 – Agreements, Off-Shore Procurement and Other Defense Services**
vi. **ITAR Part 125 – Licenses for the Export of Technical Data and Classified Defense Articles**
vii. **ITAR Part 126 – General Policies and Provisions**
viii. **ITAR Part 127 – Violations and Penalties**
ix. **ITAR Part 128 – Administrative Procedures**
x. **ITAR Part 129 – Registration and Licensing of Brokers**
xi. **ITAR Part 130 – Political Contributions, Fees, and Commissions**

- International Traffic in Arms Regulations (ITAR), 22 CFR 120-130, U.S. Department of State.
- Export Administration Regulations (EAR), 15 CFR 730-774, U.S. Department of Commerce.
- Bureau of Industry and Security (BIS), U.S. Department of Commerce.
- International Traffic in Arms Regulations (ITAR), 22 C.F.R. Parts 120-130 (2021)
- Export Administration Regulations (EAR), 15 C.F.R. Parts 730-774 (2021)
- Department of State Directorate of Defense Trade Controls, ITAR Compliance Handbook (2021)
- Bureau of Industry and Security, Compliance Guidance for Exporters (2021)
- U.S. Department of Commerce, A Guide to U.S. Export Controls (2021)
- International Traffic in Arms Regulations (ITAR), 22 C.F.R. Parts 120-130 (2021)
- Export Administration Regulations (EAR), 15 C.F.R. Parts 730-774 (2021)
- Department of State Directorate of Defense Trade Controls, ITAR Compliance Handbook (2021)
- Bureau of Industry and Security, Compliance Guidance for Exporters (2021)
- Department of State Directorate of Defense Trade Controls, ITAR Compliance Handbook (2021)
- Bureau of Industry and Security, Compliance Guidance for Exporters (2021)
- U.S. Department of Commerce, A Guide to U.S. Export Controls (2021)
- International Traffic in Arms Regulations (ITAR), 22 C.F.R. Parts 120-130 (2021)
- Export Administration Regulations (EAR), 15 C.F.R. Parts 730-774 (2021)
- Department of State Directorate of Defense Trade Controls, ITAR Compliance Handbook (2021)

- Bureau of Industry and Security, Compliance Guidance for Exporters (2021)
- U.S. Department of Commerce, A Guide to U.S. Export Controls (2021)
- International Traffic in Arms Regulations (ITAR), 22 C.F.R. Parts 120-130 (2021)
- Export Administration Regulations (EAR), 15 C.F.R. Parts 730-774 (2021)
- Department of State Directorate of Defense Trade Controls, ITAR Compliance Handbook (2021)
- Bureau of Industry and Security, Compliance Guidance for Exporters (2021)
- U.S. Department of Commerce, A Guide to U.S. Export Controls (2021)
- International Traffic in Arms Regulations (ITAR), 22 C.F.R. Parts 120-130 (2021)
- Export Administration Regulations (EAR), 15 C.F.R. Parts 730-774 (2021)
- Department of State Directorate of Defense Trade Controls, ITAR Compliance Handbook (2021)
- Bureau of Industry and Security, Compliance Guidance for Exporters (2021)
- U.S. Department of Commerce, A Guide to U.S. Export Controls (2021)
- International Traffic in Arms Regulations (ITAR), 22 C.F.R. §§ 120-130 (2021).
- Best Practices for ITAR Compliance, Export.gov (2021).
- U.S. Department of State Directorate of Defense Trade Controls. (2021). ITAR Penalties. Retrieved from https://www.pmddtc.state.gov/compliance/itar_penalties.html
- U.S. Department of Commerce, Bureau of Industry and Security. (2021). ITAR Violations and Penalties. Retrieved from https://www.bis.doc.gov/index.php/policy-guidance/guidance/violations-penalties-and-settlements/itar-violations-and-penalties

- International Traffic in Arms Regulations (ITAR), 22 CFR Parts 120-130, available at https://www.pmddtc.state.gov/regulations_laws/itar.html
- U.S. Department of State, Directorate of Defense Trade Controls, ITAR Overview, available at https://www.pmddtc.state.gov/compliance/itar_overview.html
- U.S. Department of State, Directorate of Defense Trade Controls, ITAR Compliance, available at https://www.pmddtc.state.gov/compliance/compliance.html
- International Traffic in Arms Regulations (ITAR), 22 CFR Parts 120-130, available at https://www.pmddtc.state.gov/regulations_laws/itar.html
- U.S. Department of State, Directorate of Defense Trade Controls, ITAR Violations and Penalties, available at https://www.pmddtc.state.gov/compliance/violations_penalties.html
- International Traffic in Arms Regulations (ITAR), 22 CFR Parts 120-130, available at https://www.pmddtc.state.gov/regulations_laws/itar.html
- International Traffic in Arms Regulations (ITAR), 22 CFR Parts 120-130, available at https://www.pmddtc.state.gov/regulations_laws/itar.html
- International Traffic in Arms Regulations (ITAR), 22 CFR Parts 120-130, available at https://www.pmddtc.state.gov/regulations_laws/itar.html
- U.S. Munitions List (USML), 22 CFR Part 121, available at https://www.pmddtc.state.gov/regulations_laws/itar.html
- U.S. Department of State, Directorate of Defense Trade Controls, available at https://www.pmddtc.state

Step-by-Step Directions to DDTC (Defense Trade Controls) Registration

1. Determine your company's eligibility for DDTC registration. To be eligible, your company must be engaged in the manufacture, export, or import of defense articles and related technical data.
2. Gather all necessary information and documents. This includes your company's name, address, tax ID number, business structure, and information about its ownership and management.
3. Access the DDTC's online registration portal, the D-Trade System. If you don't have an account, you'll need to create one by providing basic information about your company and your role within the company.
4. Complete the DDTC registration application form. The form will ask for detailed information about your company, including its structure, ownership, management, and activities.
5. Upload any required supporting documents. This may include your company's articles of incorporation, tax ID number, or other proof of business formation.
6. Review and submit the application. You'll be given an opportunity to review your application and make any necessary changes before submitting it to DDTC.
7. Wait for the DDTC to process your application. The processing time can vary, but typically takes several weeks. DDTC may ask for additional information or clarification if needed.
8. Receive your DDTC registration number. Once your application is approved, DDTC will assign you a registration number, which you'll need to include in all future export license applications.
9. Stay current with your DDTC registration. You'll need to renew your DDTC registration every three years and report any changes to your company's information promptly.

10. It's important to thoroughly understand DDTC regulations and the steps involved in the DDTC registration process to ensure compliance with ITAR and export control requirements.

Quick ITAR Checklist

1. Identify USML Status

Assess whether your product or service offered falls under the jurisdiction of the State Department as listed on the USML, and thus is subject to ITAR compliance requirements.

2. Understanding ITAR Components

Examine the ITAR regulations thoroughly to gain a comprehensive understanding of its multiple facets. The ITAR regulations are organized into 11 distinct sections:

ITAR Part 120 – Purpose and Definitions

ITAR Part 121 – The United States Munitions List

ITAR Part 122 – Registration of Manufacturers and Exporters

ITAR Part 123 – Licenses for the Export of Defense Articles

ITAR Part 124 – Agreements, Off-Shore Procurement and Other Defense Services

ITAR Part 125 – Licenses for the Export of Technical Data and Classified Defense Articles

ITAR Part 126 – General Policies and Provisions

ITAR Part 127 – Violations and Penalties

ITAR Part 128 – Administrative Procedures

ITAR Part 129 – Registration and Licensing of Brokers

ITAR Part 130 – Political Contributions, Fees, and Commissions

3. DDTC Registration

Complete the registration process with the Directorate of Defense Trade Control (DDTC) to be in compliance with regulations and to contribute to the advancement of US national security through the export of defense articles, services, and technologies.

4. Classification

Categorize your offerings into the proper classifications as outlined in the United States Munitions List (USML).

5. End Use

Verify the intended end-users and the intended use of your products to ensure compliance with ITAR regulations.

6. US State Department Application

Submit a request for an export license through the US State Department for the purpose of exporting defense articles, services, and technologies.

7. Reporting & Documentation

Comply with all ITAR-mandated reporting requirements by keeping accurate and organized records that can be readily accessed and inspected by the DDTC upon request.

8. Develop Export Compliance Program

Develop and maintain a comprehensive Export Compliance Program (ECP) to ensure adherence to the regulations of the Defense Trade Controls Compliance Office. This program should be tailored to the specific operations of your business and take into account potential areas of risk. It is recommended to document the ECP in writing, regularly review and update it as necessary, and have it strictly enforced by a designated compliance manager within your company.

ITAR and Export Control Compliance Checklist

- ✓ Understanding the purpose of ITAR and its regulations
- ✓ Defining ITAR-controlled items, services, and technology
- ✓ Familiarizing with the ITAR licensing requirements
- ✓ Knowing the ITAR recordkeeping and reporting requirements

I. Developing an ITAR Compliance Program

- ✓ Establishing ITAR compliance obligations and responsibilities
- ✓ Assigning a dedicated ITAR compliance officer
- ✓ Developing internal policies and procedures for ITAR compliance
- ✓ Implementing training programs for employees on ITAR regulations
- ✓ Evaluating and updating ITAR compliance strategies on a regular basis

II. ITAR and Export Control Compliance Obligations

- ✓ Identifying ITAR-controlled items and services
- ✓ Determining the ITAR licensing requirements for each item/service
- ✓ Evaluating and managing export control risks
- ✓ Developing ITAR and export control compliance strategies
- ✓ Ensuring proper recordkeeping and documentation

III. Best Practices for ITAR and Export Control Compliance

- ✓ Providing regular training and awareness programs for employees
- ✓ Implementing internal controls and processes to ensure compliance
- ✓ Developing and maintaining proper records and documentation

- ✓ Establishing relationships with business partners and customers that comply with ITAR regulations
- ✓ Regularly reviewing and updating ITAR and export control compliance strategies

IV. Penalties for ITAR Non-Compliance

- ✓ Understanding the consequences of ITAR non-compliance
- ✓ Being aware of the potential penalties for ITAR violations
- ✓ Implementing effective controls to prevent ITAR non-compliance

V. ITAR and Export Control Compliance Monitoring and Auditing

- ✓ Regularly monitoring compliance with ITAR regulations
- ✓ Conducting periodic internal audits to assess compliance
- ✓ Implementing corrective actions to address any compliance gaps.

Made in the USA
Middletown, DE
02 November 2024